Fossil Fuels

Carmel Reilly

NELSON
CENGAGE Learning™

Australia • Brazil • Japan • Korea • Mexico • Singapore • Spain • United Kingdom • United States

Fossil Fuels

Fast Forward
Turquoise Level 17

Text: Carmel Reilly
Editor: Cameron Macintosh
Design: Stella Vassiliou
Series design: James Lowe
Production controller: Seona Galbally
Photo research: Gillian Cardinal
Audio recordings: Juliet Hill, Picture Start
Spoken by: Matthew King and Abbe Holmes
Reprint: Jennifer Foo

Acknowledgements
The author and publisher would like to acknowledge permission to reproduce material from the following sources: Photographs by AAP Image/AFP Photo/Mustafa Ozer, p15 bottom, 23 centre top; AAP Image/AFP Photo/Torsten Blackwood, p15 top, 23 top; AAP Image/AFP Photo/Toru Yamanaka, p14, 23 bottom, back cover; AAP Image/EPA/Paul Hilton, p 22-23; Alamy/Imagine That, p5 bottom; Alamy/Phototake Inc./Tom Carroll, p 6; iStockphoto.com/Jill Fromer, p11 bottom right; iStockphoto.com/ Louis Aguinaldo, p11 centre bottom; iStockphoto.com/Valentin Essen, p11 bottom left; Photo Edit/David Frazier, p19; Photolibrary/Age Fotostock/Paul Gibbard, p7; Photolibrary/Images.com/Scott McDermott, p21; Photolibrary/Imagestate Ltd/Grant Frank, p5 top; Photolibrary/Index Stock Imagery/Spence Inga, front cover inset left; Photolibrary/Oxford Scientific Films/Chris Sharp, p16; Photolibrary/Peter Arnold Images Inc./Jeri Gleiter, p10; Photolibrary/Photo Researchers Inc./Georgia Lowell, p12; Photolibrary/Science Photo Library/Martin Bond, p20 left, 20 right; Photolibrary/Science Photo Library/Alfred Pasieka, p8; Photolibrary/Science Photo Library/Gary Hincks, p9; Photolibrary/Science Photo Library/Mauro Fermariello, p13; Photolibrary/Workbook, Inc./Ganci Carson, pp front cover

ISBN 978 0 17 012631 1
ISBN 978 0 17 012621 2 (set)

Cengage Learning Australia
Level 7, 80 Dorcas Street
South Melbourne, Victoria Australia 3205
Phone: 1300 790 853

Cengage Learning New Zealand
Unit 4B Rosedale Office Park
331 Rosedale Road, Albany, North Shore NZ 0632
Phone: 0508 635 766

For learning solutions, visit cengage.com.au

Printed in Australia by Ligare Pty Ltd
7 8 9 10 11 12 13 21 20 19 18 17

THE UNIVERSITY OF MELBOURNE

Evaluated in independent research by staff from the Department of Language, Literacy and Arts Education at the University of Melbourne.

Fossil Fuels

Carmel Reilly

Contents

Chapter 1

THE PROBLEM WITH FOSSIL FUELS

Oil, natural gas and coal are all **fossil fuels**. Between them, they provide more than 80 per cent of the energy used in the world today.

an oil rig

an open cut coal mine

burning natural gas in the kitchen

However, there are two major problems with the use of fossil fuels.

First, fossil fuels will not last forever. At some point they will run out.

Second, when fossil fuels are burnt for energy, they cause pollution, which is bad for the environment.

While most people agree that we need to do something about these problems, different people have different ideas about what to do.

Chapter 2

A CLOSER LOOK AT FOSSIL FUELS

It is believed that oil, natural gas and coal were formed from the remains – or fossils – of plants and tiny animals that lived about 300 million years ago (before the time of the dinosaurs).

the fossil of an extinct shell fish

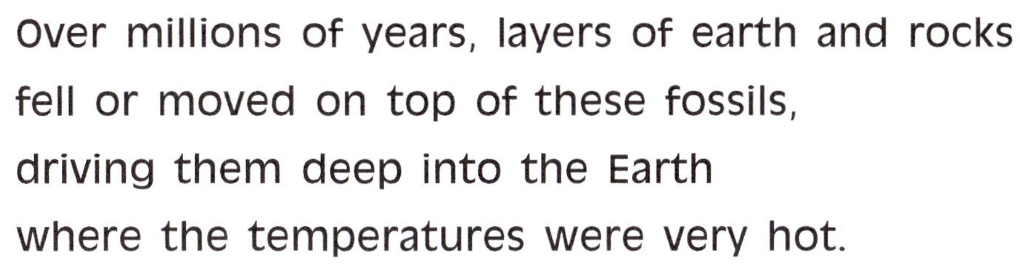

Over millions of years, layers of earth and rocks fell or moved on top of these fossils, driving them deep into the Earth where the temperatures were very hot.

Over a long time, huge chemical changes happened to them.

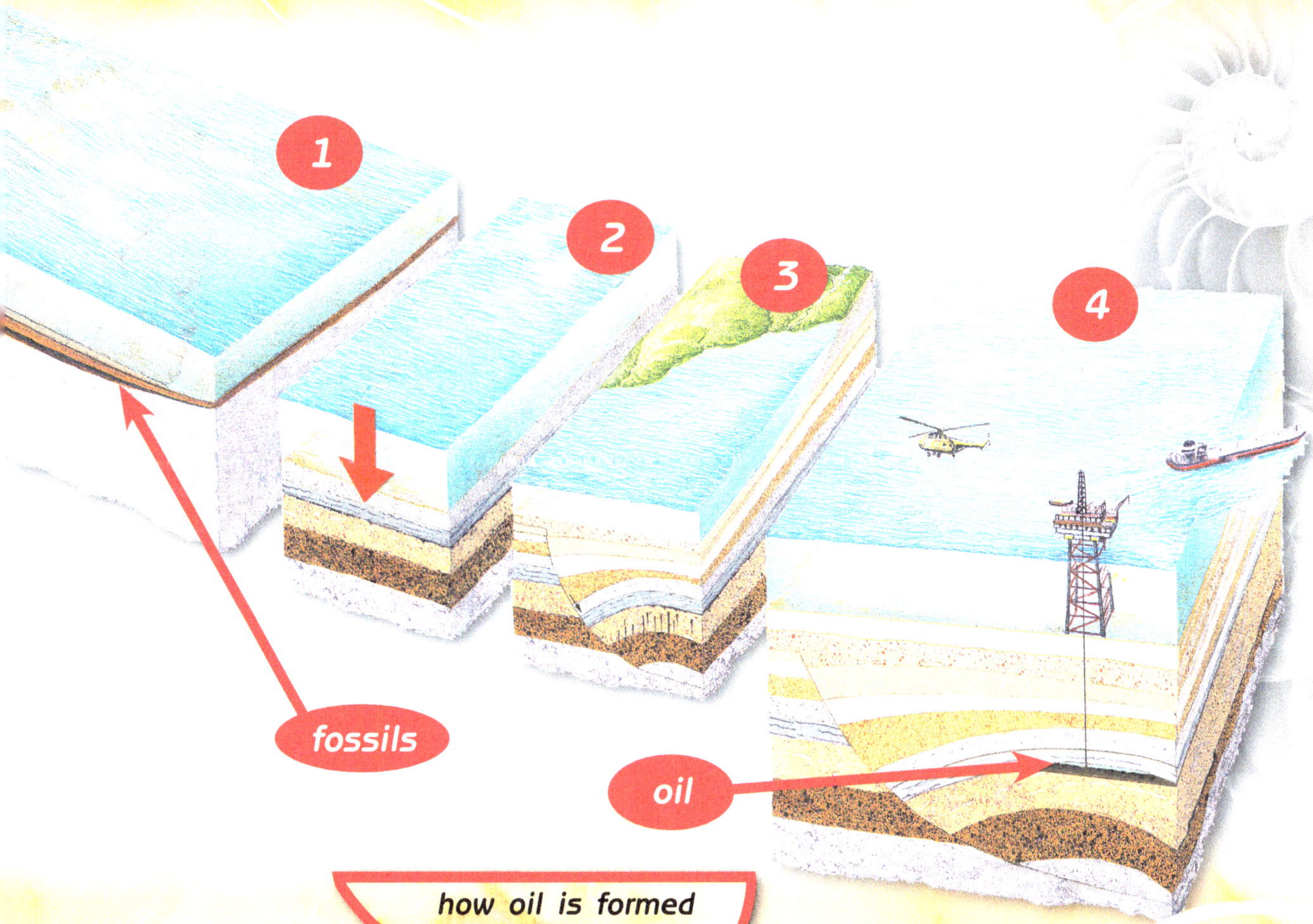

how oil is formed

Today, we use fossil fuels to provide all kinds of energy.

Oil is made into fuels that power cars and other kinds of transport.

coal power plant

Coal and natural gas are mostly used as a fuel to make electricity.

All three are used for cooking, and for warming our homes and workplaces.

On top of that, oil and natural gas can also be mixed with other things to make chemicals and plastics.

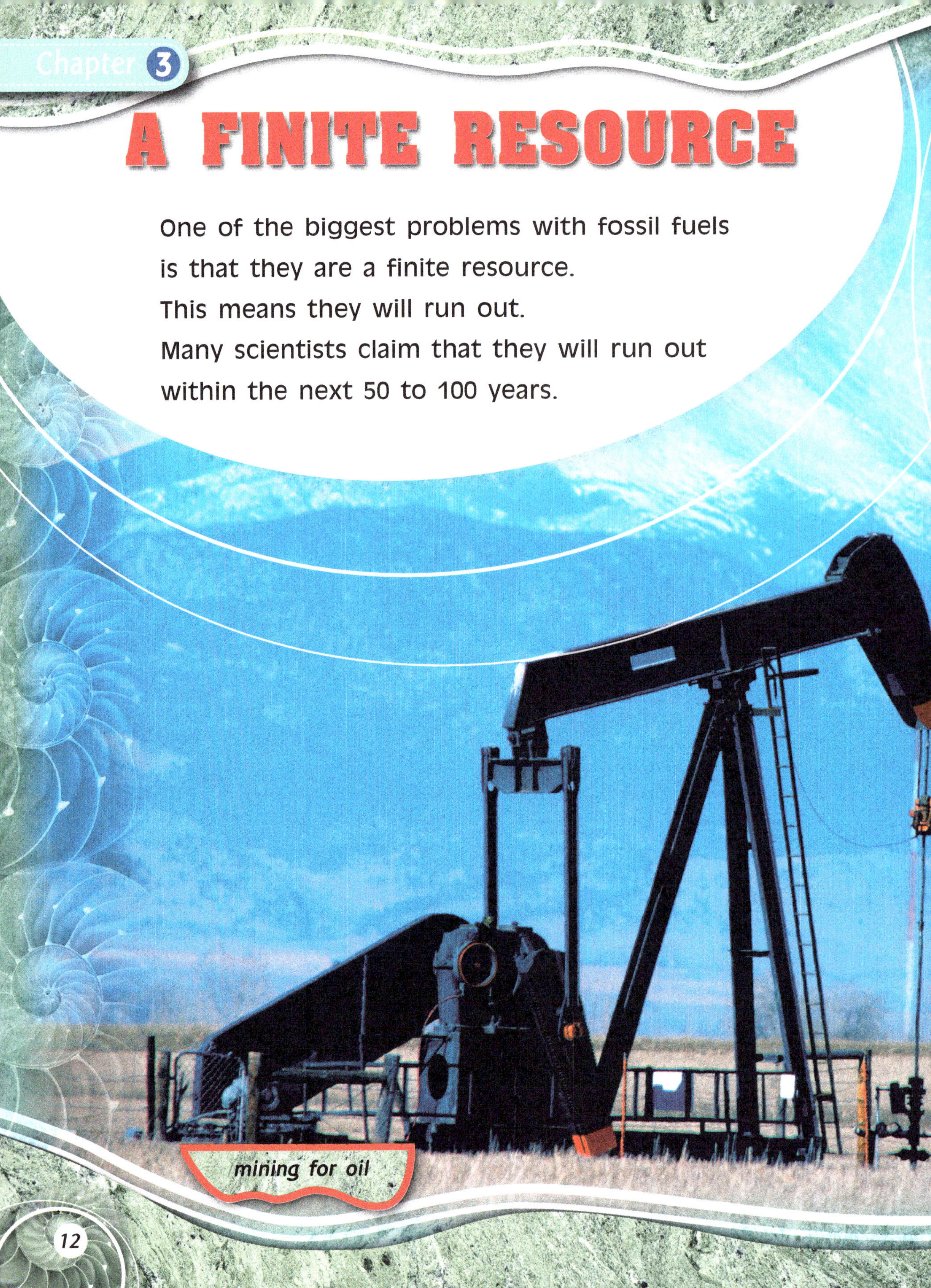

Chapter 3

A FINITE RESOURCE

One of the biggest problems with fossil fuels is that they are a finite resource.
This means they will run out.
Many scientists claim that they will run out within the next 50 to 100 years.

mining for oil

However, people who support the use of fossil fuels argue that a lot can be done to make them last longer.

For example, when oil is taken out of the earth, a lot is left behind.
Some scientists are looking at better ways to reach this oil.
If they can do this, there will be more oil for people to use.

Chapter 4

THE RIGHT THING TO DO

Even if there is more oil to use, should it be used up? Oil is still a finite resource, which means it will run out. This means that, sooner or later, we will need to find other kinds of energy for transport.

electric car

solar and wind power ferry

solar racing car

Chapter 5

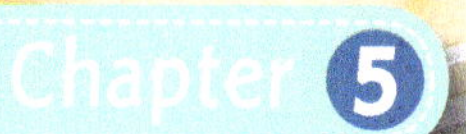

THE ENVIRONMENT

Burning so much fossil fuel has made a lot of pollution. In turn, this pollution has caused changes to the environment, such as **global warming**.

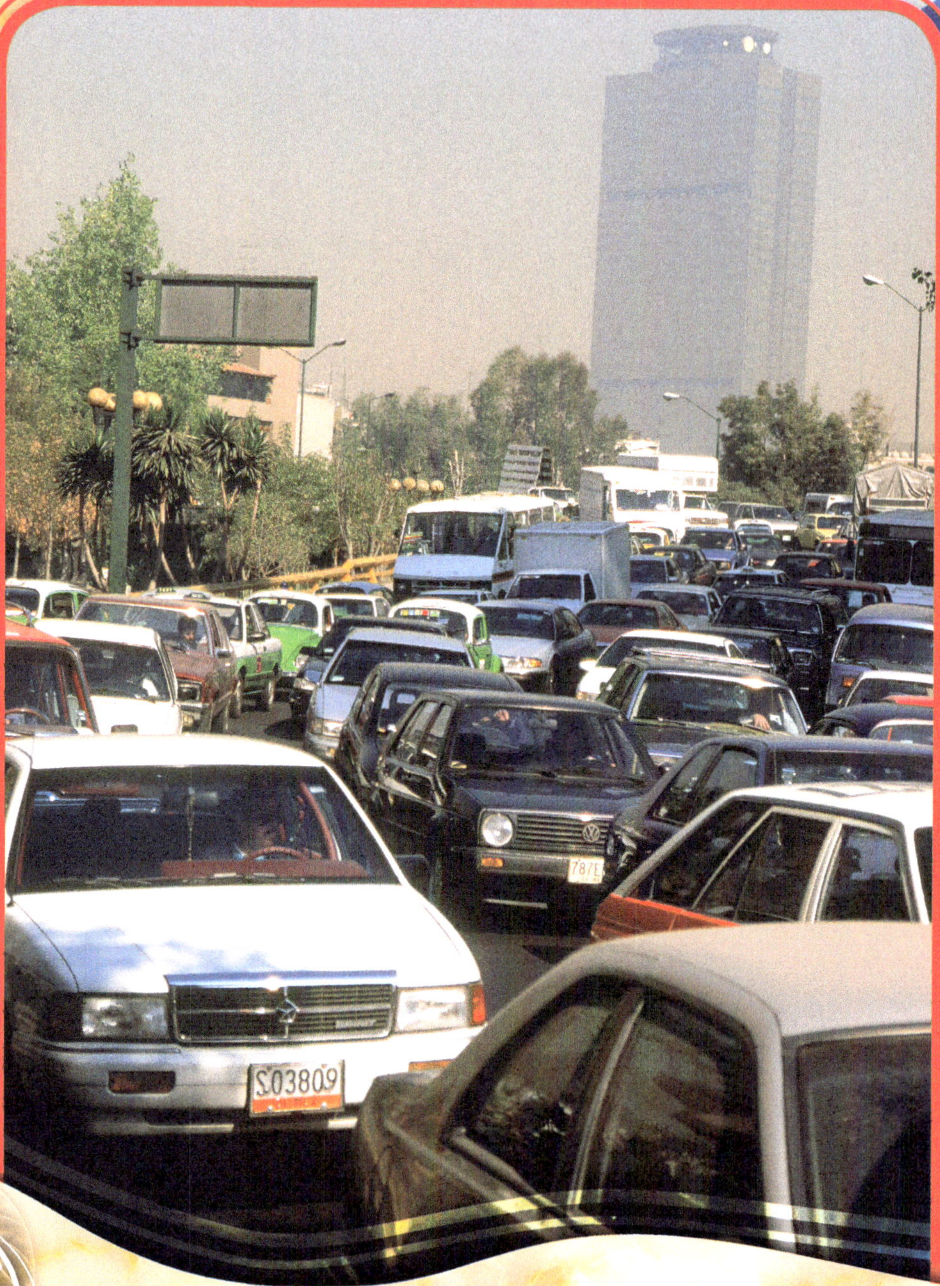

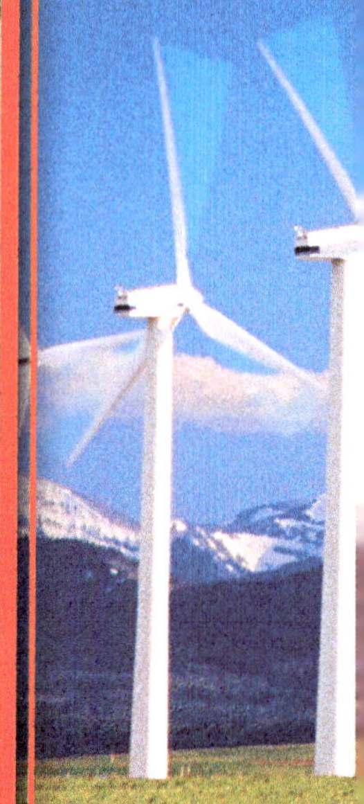

Because of this, many people argue that we should find other ways to run transport and make electricity, rather than trying to extend the use of fossil fuels.

People who support the use of fossil fuels claim that they are the best form of energy at the moment.

They say that scientists are working on ways to stop fossil fuels making as much pollution, which will be better for the environment.

They also argue that other forms of power, like **nuclear power**, are not safe for the environment either.

As yet, solar, water and wind power cannot provide all the energy the world needs.

People who are against the use of fossil fuels say that we need to spend more money on researching energy forms that do not cause as many problems for the environment. These energy forms include solar, water and wind power.

water power electricity plant

a solar house

These people also argue that we should look at ways to reduce our energy use so that we don't require as much of any kind of fuel.

Chapter 6

THE BIG QUESTION

Most people would agree that we need to reduce or change the way we use fossil fuels as soon as we can.

The big question is: what is the best way to do this?

Glossary

fossil fuels fuels such as gas or coal, formed from the remains of living things

global warming the gradual rise in the Earth's temperature

nuclear power power generated by a nuclear reactor

Index